IMAGES
of America

NAVAL AIR STATION, LAKEHURST

ZR-3, the USS *Los Angeles*, is seen on the high mast at Naval Air Station, Lakehurst in 1927.

IMAGES
of America

NAVAL AIR STATION, LAKEHURST

Kevin Pace, Ronald Montgomery, and Rick Zitarosa

ARCADIA
PUBLISHING

Published by Arcadia Publishing
Charleston, South Carolina

Library of Congress Catalog Card Number: 2002116263

For all general information contact Arcadia Publishing at:
Telephone 843-853-2070
Fax 843-853-0044
E-mail sales@arcadiapublishing.com
For customer service and orders:
Toll-Free 1-888-313-2665

Visit us on the Internet at www.arcadiapublishing.com

ZR-3, the USS *Los Angeles*, and ZRS-4, the new USS *Akron*, are pictured in Hangar No. 1 at Lakehurst in 1931.

CONTENTS

ACKNOWLEDGMENTS

Many have contributed to this work, some of them long gone and many years ago.

Michael C. Miller and H.J. "Hank" Applegate sowed the seeds of historic preservation at the old Naval Air Station, Lakehurst over a quarter century ago.

Among those who have since passed away, we wish to thank Capt. Frederick N. Klein, Lt. Comdr. James M. Punderson, Lt. Comdr. Leonard "Gene" Schellberg, Comdr. Charles Mills, Capt. M. Henry Eppes, Aviation Chief Machinist's Mate Julius "Gene" Malak, Dr. Douglas H. Robinson, and Vice Adm. Charles E. Rosendahl.

John Iannaccone (of the USS *Los Angeles*) and John Lust (of the USS *Akron*), two of the very last original sky sailors left in existence, continue to share themselves unselfishly with the Navy Lakehurst Historical Society. Happily, Charlie Bish, Morris "Mac" McConnell, C.C. Moore, Jim Shock, Hepburn Walker Jr., Lt. J. Gordon Vaeth, Comdr. Lundi Moore, and Capt. John Kane are all still with us today.

Capt. Mike Dougherty, Capt. Steve Himes, Capt. Dwight Cousins, Capt. Mark Bathrick, and staff members have been most gracious hosts and are proof that Navy Lakehurst remains in good hands today.

Lastly, we would like to thank the officers and volunteer staff of the Navy Lakehurst Historical Society for their support and encouragement in this work, particularly Alex Swaim, who spent so many days and nights working on this project with us, and Carl Jablonski, President.

INTRODUCTION

Lakehurst's history is that of airships in the U.S. Navy, but many other aspects of the base are largely unknown, including its background as an army chemical warfare proving ground and its later roles of manufacture, testing, and training in support of naval aviation.

Continuously evolving as a vital component of national defense, Lakehurst has seen its share or triumphs and setbacks over its eight decades of operation, and this is expected to continue as the military redefines itself in a fast-paced high-tech era. Despite several proposals to consolidate or eliminate Lakehurst's role over the years, the place continues to flourish, like a cat with nine lives.

This book has been a labor of love for the authors, all of whom have a deep interest in the history of Navy Lakehurst. An enjoyable project for us all, it has enabled us to make use of our collective interests, research, and the extensive archival and photographic material of the Navy Lakehurst Historical Society, founded in 1986.

We look forward to many more years of celebrating the rich past and the exciting future achievements in aviation, community, and government service at this place so dear in our hearts.

—Kevin Pace, Ronald Montgomery, and Rick Zitarosa
Navy Lakehurst Historical Society
www.nlhs.com

One

THE LAKEHURST
PROVING GROUNDS

Beyond the main gate at Camp Kendrick is the army's principal experimentation site for chemical warfare technology. The presence of a military installation did much to improve the economy of the region. Even with live rounds of mustard gas and other hazardous chemicals being tested here, inhabitants of nearby Lakehurst (well known in its day as a "health resort") did not seem to mind.

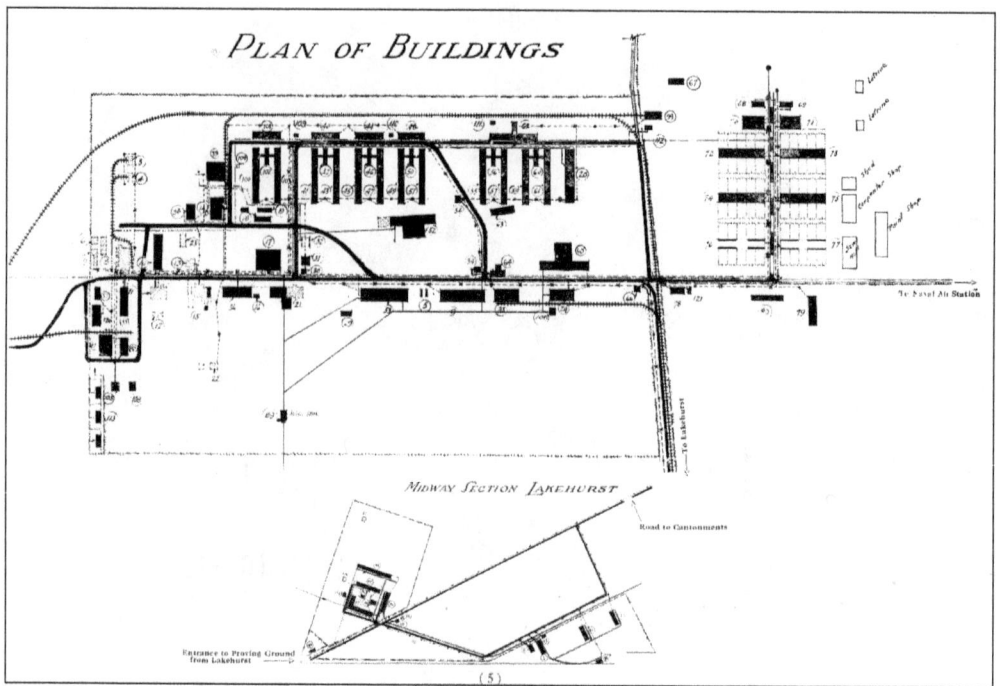

PLAN OF BUILDINGS

This 1923 map of the proving grounds property shows the extensive infrastructure that grew out of the swampy Pinelands amid America's bid to develop chemical warfare technology in World War I.

Looking west, this view shows the main street of the Lakehurst Proving Grounds in February 1918. Note the utility poles and the standardized semiportable buildings, most of which were later sold off and relocated to various sites around Ocean County, where many continue to stand today. The site shown is now heavily overgrown, with only a few foundation footings to mark the location of the once busy base.

Two unsmiling workers of the Eddystone Ammunition Corporation, a subsidiary of the Baldwin Locomotive Works, pose with "company transportation." Eddystone's venture into arms manufacture was the principal force behind the acquisition of the Lakehurst site as a munitions test range in 1915.

The completed headquarters building is pictured in 1917.

In this 1916 view, men test-fire Eddystone shells made for the Russian government. Despite imperial Russian army inspectors stationed at Lakehurst between 1915 and 1917, there were reports that the shells shipped for use overseas were of only fair quality. However, with the United States soon to be directly involved in World War I, the Lakehurst Proving Grounds soon came under U.S. Army control.

Pictured is the wood observation tower for Trenches No. 1 and No. 2. Smoke bombs and fires were set in the distance, and when a favorable wind direction was indicated, gas shells were shot off and their effects were recorded. Soldiers with appropriate respirator gear would retrieve sample flasks and record the condition of test animals in trenches.

These range officers, pictured at the Lakehurst Proving Grounds in 1918, are equipped with the latest type of steel helmets and gas masks.

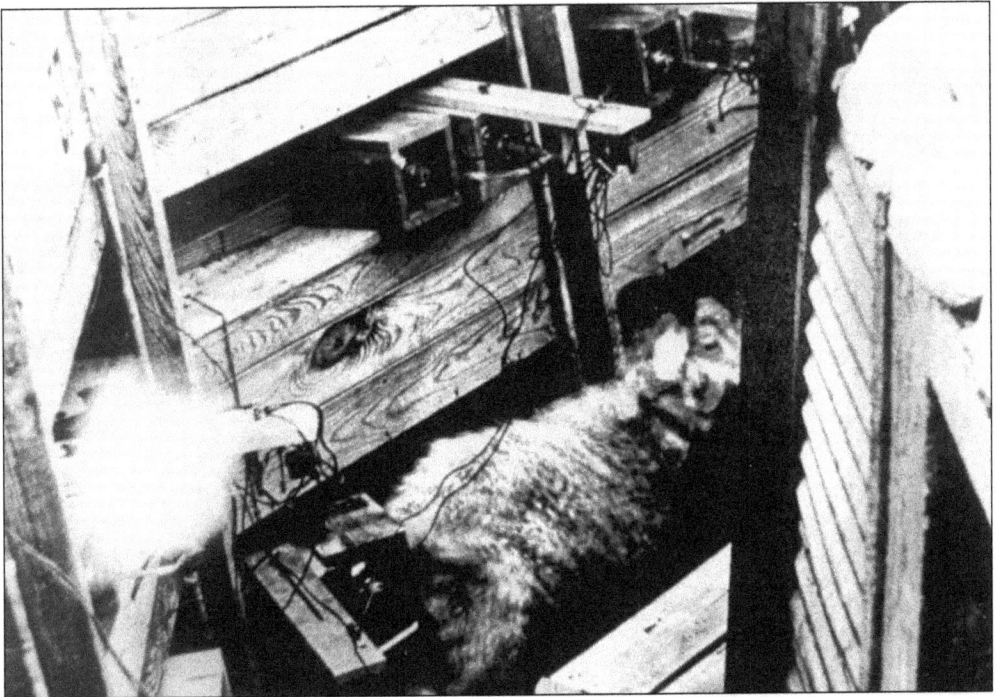

Sheep are set up for monitoring in the test trenches. Note the sampling flasks placed in midlevel shelving. Testing the effects of deadly poison gases in various concentrations under controlled conditions, the proving grounds contributed immeasurably to advances in chemical warfare.

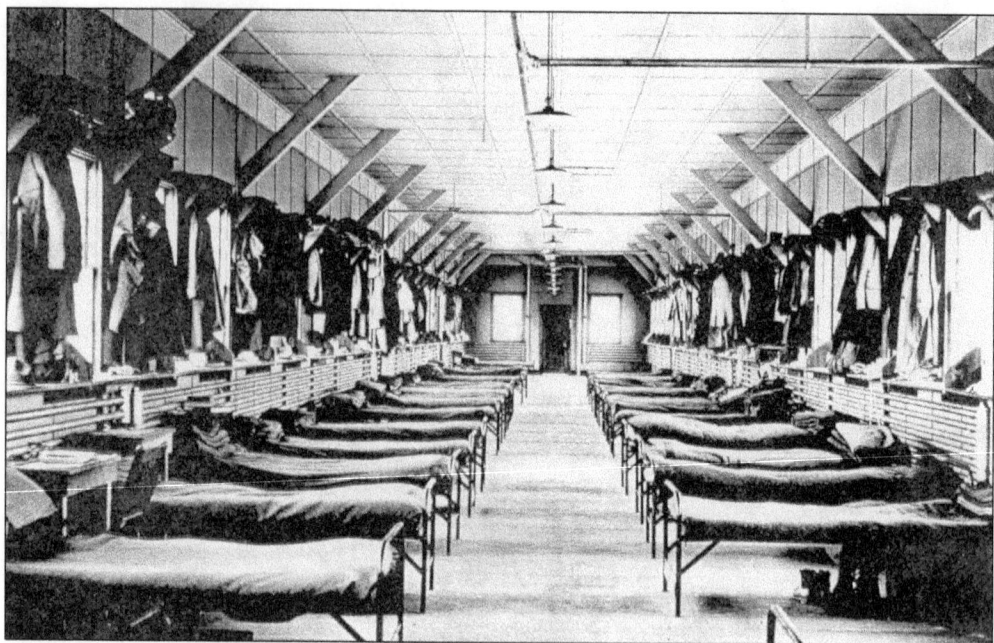

In contrast to the doughboys in the rotting trenches of Europe, soldiers stationed at Lakehurst enjoyed steam heat, electric lights, and "thoroughly modern latrine facilities." This view shows the interior of a barracks.

A narrow-gauge 12-ton geared steam locomotive was used to haul supplies and munitions around the proving grounds. Heavy deliveries of supplies, troops, and guns came via a Central Railroad of New Jersey connection just north of the town of Lakehurst.

Pictured is the junction point where a spur from the Central Railroad of New Jersey interchanged with the narrow-gauge railway serving the army proving grounds. Sandy, primitive roads necessitated the use of trains to move heavy ordnance and bulk items.

This 1918 view looks north from the trenches of the proving grounds. Located near the test facility and runway areas far out on the base, this was the site of the first full-scale gas warfare experiments in the United States.

Used for throwing barbed-wire projectiles, this gun was tested at Lakehurst between 1915 and early 1917, the period when the site was controlled by the Eddystone Ammunition Corporation.

Col. William S. Bacon, commanding officer, tours the base for the benefit of visiting Maj. H.R. LeSueur of the British Military Mission.

Two

A Home for the Navy Lighter-Than-Air Program

THE BUILDINGS AND FIXTURES
OF THE
LAKEHURST PROVING GROUNDS
LAKEHURST, NEW JERSEY
WILL BE SOLD AT
PUBLIC AUCTION SALE
By Order of the Secretary of War

OPEN
FOR
INSPECTION

120
FRAME
BUILDINGS

BIRD'S EYE VIEW

TUESDAY, MARCH 27th, 1923, at 12 o'clock noon
On the premises of the Proving Grounds, Lakehurst, N. J.
FOR FULL PARTICULARS APPLY TO QUARTERMASTER SUPPLY OFFICER, 1st AVENUE AND 59th STREET, BROOKLYN, N. Y.
SAMUEL T. FREEMAN & CO., Auctioneers, 1519-1521 Chestnut Street, Philadelphia, Pa.

With the cessation of hostilities, ambitious plans for further experiments at Camp Kendrick were shelved. The base was closed, and the buildings were put up for sale. In the summer of 1919, Lt. Comdr. Lewis Maxfield, a skilled, enthusiastic navy dirigible pilot, recommended Lakehurst as the new home for the navy's lighter-than-air (LTA) program.

As acting secretary of the navy, Franklin D. Roosevelt initiated the deal to acquire the original 7,400-acre (later expanded) Lakehurst tract from the army for use as a "dirigible field." The cost came to $13,088, and the deal was finalized by summer 1919. The U.S. Navy Department of Yards and Docks began drawing up specifications for a giant hangar and support facilities.

In August 1919, the Lord Construction Company of Philadelphia won the hangar bid at a price of $2.9 million. The job of transforming the former army proving ground into a naval air station began immediately. A good water supply, rail connections, plentiful local labor, and the ability to use some of the army buildings already on site kept the project moving along.

As the tradesmen worked steadily toward the completion of what was (for a time) the largest open interior space in the world, components for Fleet Airship No. 1 (later Rigid Airship ZR-1) were being manufactured at the Naval Aircraft Factory, Philadelphia, while ZR-2 was being built for the U.S. Navy by the British Royal Airship Works at Cardington, England. It was expected that the British-built ship would make its transatlantic flight by late summer in 1921 and that erection of ZR-1 components would commence around the same time.

As the U.S. Navy originally planned to use hydrogen gas in its airships, the facilities of Hangar No. 1 and the base were appropriately completed. The hangar featured extensive gastight electrical service, as well as fireproofing and ventilation technology. The station gas plant was designed to manufacture 75,000 cubic feet of hydrogen per day.

Consisting of a 10-arch steel framework sheathed with asbestos, the new Lakehurst hangar grew rapidly from the concrete footings poured in the sandy soil. By January 1921, the structure was essentially complete. The four 1,300-ton electrically operated doors were in working order, and finishing touches were being applied when Capt. Frank T. Evans put Naval Air Station, Lakehurst into commission on June 26, 1921.

The base soon outgrew the original wooden firehouse built in 1921. A modern, WPA-style brick building was completed as a replacement in 1935, before the base's greatest fire, the *Hindenburg* disaster of May 6, 1937. Busy air station firefighters have responded to their share of blazes both on and off base over the years, particularly when forest fires threaten neighboring Pine Barrens communities. The brick firehouse is still in use today, and the old wooden firehouse exists as a workshop.

This *c.* 1924 view looks north from the roof of Hangar No. 1. The building in the right background is the balloon hangar, featuring a canvas door that could easily blow out (rather than the building blowing apart) in the event of hydrogen fire or explosion. With scarce helium reserved for use in blimps and rigid airships, training and kite balloons still used hydrogen and were appropriately quarantined there. The helium plant and powerhouse are on the left.

Airship Hangar, Lakehurst, N.J.

Hangar No. 1 was the first permanent address of the navy's lighter-than-air operations. Its dimensions are 966 feet (overall length), 807 feet (interior length), 263 feet (interior door width), 224 feet (overall height), and 178 feet (interior height, floor to arch). The hangar remained basically unchanged until the mid-1980s, when camouflage pattern asbestos siding panels were replaced with aluminum. Also at that time, the original amber-tinted glass (designed to minimize solar effects on helium temperatures and delicate airship fabrics) was replaced with a safety-glass substitute. The building was declared a National Historic Landmark in 1968.

Lakehurst, N.J. Naval Air Station
Married Officers Quarters

Officers' housing eventually grew to include substantial brick homes and bachelor officers' quarters, constructed by the early 1930s. They still serve their intended purpose, having outlasted wooden buildings that were added during the 1942–1944 expansion. Today, buildings on the base are either very new or very old, with little in between.

Three

LIGHTER-THAN-AIR
ACTIVITIES IN THE 1920S

The Allies received several Zeppelins as spoils of war from Germany and also built several based on the designs of captured German models. One such copy, the British R-34, made the first round-trip crossing of the Atlantic by air between England and Mineola, Long Island, in July 1919. The vessel's speed was hardly anything to brag about (108 hours westbound, 75 hours eastbound), but the flight was convincing proof that the big rigid airships had serious long-range capability. The U.S. Navy was soon negotiating with the British to build a rigid airship and to train an American crew.

More than 100 Zeppelin-type airships carried out long-range bombing and scouting missions for the German army and navy during World War I. By mid-1916, high-flying Allied planes firing incendiary and explosive machine gun ammunition turned the tide against the "Zeppelin menace." Although Zeppelins were known to be big, slow, and lifted aloft by flammable hydrogen gas, the major Allied powers sought to develop the big rigid airships for military and commercial purposes after the war.

The erection of the first American rigid airship began in early 1922. It was based largely on the plans of the captured World War I German navy Zeppelin L-49. With the ZR-2 tragedy in mind, U.S. Navy designers spent months rechecking their design stress calculations for ZR-1. In this view, ZR-1 is nearing completion in Hangar No. 1 in the spring of 1923. A Central Railroad of New Jersey camelback freight locomotive has just delivered another carload of components from the Naval Aircraft Factory, Philadelphia. For years, special trains brought sightseers to watch the comings and goings of Lakehurst's big sky ships.

24

Beginning in 1915, the U.S. Navy conducted a modest lighter-than-air program, flying nonrigid airships (or blimps). The Howden Detachment in Yorkshire, England, provided rigid airship training for navy personnel assigned to R-38 (American ZR-2). Lightly built for high-altitude flight (with high engine horsepower for speed), the new British ship's first flights were plagued with design and structural strength problems. On its fourth trial flight, August 24, 1921, the ship broke apart and fell burning into the River Humber at Hull, with 28 British and 16 Americans killed, including Lt. Comdr. Lewis Maxfield, who was designated commanding officer and was the leading U.S. Navy proponent of rigid airship development.

The ZR-1 keel area at Frame No. 60 (about 190 feet forward of the stern) is shown during construction in the spring of 1923. Note the intricate latticework girders, a 9-inch-wide plywood walkway running from stem to stern, and the nest of 113-gallon fuel tanks (some droppable as emergency ballast). Emergency rudder and elevator control wheels are not yet mounted on their stands (foreground).

Army and navy personnel are shown in spring of 1923 at the Lakehurst lighter-than-air school for rigid airships. The "professor" of the school was former German Zeppelin Airship Construction Company test pilot Anton Heinen (front row, 3rd from left). Also seen here are Comdr. Frank McCrary, first commanding officer of ZR-1, and Comdr. Ralph D. Weyerbacher of the U.S. Navy Construction Corps, the builder of ZR-1 (front row, 6th and 7th from left, respectively). Other notables in the photograph include U.S. Army Capt. William E. Kepner (middle row, far left), Lt. Charles E. Rosendahl (middle row, 11th from left), and Lt. Raymond F. Tyler (back row, 7th from left).

The ZR-1 *Shenandoah*, as completed, sails above Hangar No. 1. Its first flight was on September 4, 1923, with the christening and commissioning on October 10. Some 680 feet long, 91 feet high, and capable of moving through the air at 60 miles per hour, ZR-1 was the biggest thing that had ever flown in the United States. It was also an enormous technical challenge, and navy personnel had to learn the art of flying the giant rigids.

In November 1923, the *Shenandoah* began experiments with the 160-foot mooring mast on Lakehurst's west field. Developed in Britain, mooring masts enabled airships to operate independent of expensive hangars. Early mast mooring was an adventure, as strong winds caused the ship to weave and dip, and a crew always had to be aboard to "fly" the airship at the mast.

At 6:44 p.m. on January 16, 1924, the *Shenandoah* broke free from the Lakehurst high mast in a gale. Failure of the mooring cone left a huge hole in the ship's nose. Two forward gas cells collapsed, and a large section of fabric covering tore from the top tail fin. German instructor Anton Heinen and a skeleton crew rode out the storm and brought the runaway airship back to Lakehurst after a flight of nine hours. A media event over radio and in banner headlines across the nation, the saga of this battle against the elements turned into a public relations triumph for the U.S. Navy's lighter-than-air program.

Amid $78,000 in repairs and an acrimonious "personnel shakeup," Lt. Comdr. Zachary Lansdowne (right) took command of the *Shenandoah* on February 16, 1924. Naval aviator No. 105, with wartime rigid airship training in England, Lansdowne was the first American to cross the Atlantic westbound by air, as a naval observer aboard the R-34. Lakehurst's commanding officer, Comdr. Jacob Klein (left), was among those bitterly unhappy to see an "outsider" brought in. So began a decades-long pattern of feuding among various officers and commands at Lakehurst.

Following three months of repairs, the *Shenandoah* was in the air again on May 22, 1924. The bow cone and tail fins were strengthened, and Engine No. 6 was replaced by powerful radio equipment and a tiny galley. Helium shortage was acute. Exhaust water-recovery condensers (designed to recover 122 pounds of water ballast for each 100 pounds of gasoline consumed) were fitted to three of the five engines to help keep the ship flying in equilibrium and to eliminate the need to valve off expensive helium when landing.

Off Newport, Rhode Island, on August 8, 1924, the *Shenandoah* made the first sea mooring to a mast on converted navy tanker *Patoka*. A floating mobile base, the *Patoka* was intended to help extend range and load-carrying abilities of the rigid airship at sea for operations with the fleet. Note the shiny pigment of the new nose and the water-recovery condenser above Engine No. 1.

Gen. Mason M. Patrick (center) and Col. Chalmers G. Hall (right), both of the U.S. Army Air Service, pose with Lt. Comdr. Zachary Lansdowne outside Hangar No. 1 in 1925. For many years, the army tried to horn in on the development of the rigid airships, and the rivalry was fierce at times. Aboard for the *Shenandoah*'s last flight, Hall survived to give the court of inquiry a heartfelt testimony to the dedication and bravery exhibited by Lansdowne and his crew in their time of trial and tragedy.

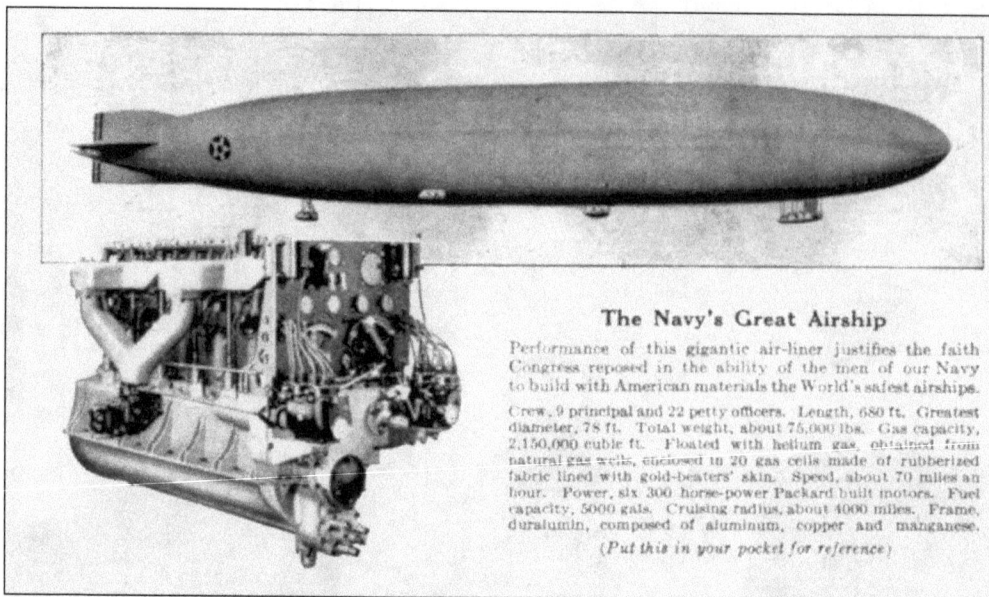

The Navy's Great Airship

Performance of this gigantic air-liner justifies the faith Congress reposed in the ability of the men of our Navy to build with American materials the World's safest airships.

Crew, 9 principal and 22 petty officers. Length, 680 ft. Greatest diameter, 78 ft. Total weight, about 75,000 lbs. Gas capacity, 2,150,000 cubic ft. Floated with helium gas, obtained from natural gas wells, enclosed in 20 gas cells made of rubberized fabric lined with gold-beaters' skin. Speed, about 70 miles an hour. Power, six 300 horse-power Packard built motors. Fuel capacity, 5000 gals. Cruising radius, about 4000 miles. Frame, duralumin, composed of aluminum, copper and manganese.

(Put this in your pocket for reference)

Publicity of the Packard Motor Company features the 1,551-cubic-inch, 300-horsepower, high-compression engines developed for the *Shenandoah*. A total of 13 of these were manufactured (six for the ship, seven spares) at $14,000 apiece. Overall, the engines gave excellent service, although failures on the ship's last flight occurred as they were pushed beyond their limits.

The *Shenandoah* is pictured moored to the seagoing mast on the *Patoka*. Note the German-style control car, originally built to hold an engine at the rear. The strut-wire attachments proved tragically inadequate in a storm over Ohio on September 3, 1925, when Lt. Comdr. Zachary Lansdowne and six others were killed after the bridge of the *Shenandoah* broke free from the disintegrating airship and fell away.

The *Shenandoah*'s engineers are pictured in the spring of 1925. Fifth from the left in the front row is Chief Warrant Machinist Shine S. Halliburton, the senior engineer. Note that one chief is wearing tennis sneakers, essential for sure footing inside the giant ship. The men chosen for the lighter-than-air program were considered the elite of naval aviation.

After 59 flights ranging from the Atlantic seaboard to the Pacific Northwest, the *Shenandoah* met a tragic end while on a publicity tour over the Midwest fair circuit on September 3, 1925. Near Ava, Ohio, a violent squall tore the airship apart with high winds and vortex air currents. Frenzied spectators descended on the shattered wreck and picked it bare. Thanks to nonflammable helium, there was no fire, but 14 out of 43 aboard were killed. Navy lighter-than-air operations spent the next few years engulfed in uncertainty and controversy.

On October 15, 1924, with the *Shenandoah* away on its 19-day transcontinental "rim flight," Naval Air Station, Lakehurst welcomed the new ZR-3 after its 81-hour flight from Germany. Months of delicate negotiations resulted in this special "compensation airship" being completed for the U.S. Navy. The new airship was promptly suspended from the hangar rafters, and its hydrogen was released. There was so little helium on hand that Lakehurst personnel had to wait for the return and lay-up of the *Shenandoah* before ZR-3's gas cells could be inflated, prompting humorist Will Rogers to quip, "The United States has two airships and only one set of helium."

ZR-3's German delivery crew stands with U.S. Navy officers in Hangar No. 1 in October 1924. Capt. George W. Steele Jr. (front row, center) made the transatlantic flight as a naval observer and became the first American commanding officer of the airship. To the right of Steele is Dr. Hugo Eckener, chairman of the German Zeppelin Airship Construction Company and commander of ZR-3 on the 5,000-mile delivery flight. Eckener's achievements with Zeppelins were front-page news for the next 15 years. Capt. Ernst Lehmann (far right), Eckener's right-hand man for many years, perished in the *Hindenburg* disaster of 1937.

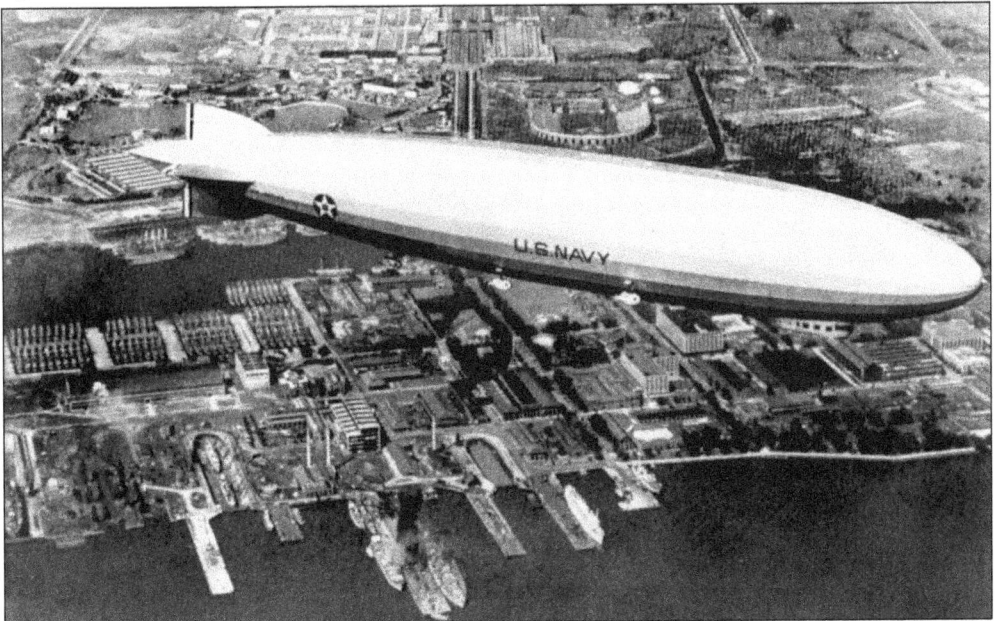

Arguably one of the most beautiful airships ever built, the *Los Angeles* shows off its classic lines as it sails above Naval Air Station, Mustin Field at the Philadelphia Navy Yard. Lt. Comdr. Charles E. Rosendahl was in command from 1926 to 1929.

In the air station motor shop, engineers gather around a German Maybach engine from the *Los Angeles* in December 1924. It took a few years before navy lighter-than-air mechanics learned all the secrets to keeping the German motors operating smoothly. Note the side inspection covers, with 10 of the 12 cylinders removed. Standing third from the right is Chief Warrant Machinist Emmet "Casey" Thurman, who served in the lighter-than-air program through World War II. On the far right is Aviation Chief Machinist's Mate William Russell, who was later lost aboard the *Akron*.

The *Los Angeles* is pictured here on the Lakehurst high mast in 1925. Note the mast elevator shaft and cable for the tail drag. Water-recovery units had not yet been installed, meaning that the airship had to valve a lot of helium in order to get heavy enough for landing after burning up fuel and provisions in flight.

On August 25, 1927, the *Los Angeles* was caught in a freak gust of wind at the high mast. Despite a tail drag, the ship performed a dramatic headstand for about four or five minutes. Miraculously, there was no serious damage to the airship or those aboard. Experiments immediately began with stub masts and taxiing cars holding the ship closer to the ground during mooring. Little used after 1928, the high mast was dismantled in 1934. The mast house at the foot of the tower survives today as Quarters W at the current Naval Air Engineering Station, Lakehurst.

This radio was set in the crew space off the keel area of the *Los Angeles*. Although it was crude by today's standards, the *Los Angeles* officers and men enjoyed hot food, comfortable bunks, and washroom and recreational space unavailable on any other type of aircraft. A very popular smoking room was added *c.* 1930. Located safely away from the gasoline tanks, it was a regular haven for those off duty.

The *Los Angeles* featured a modern electric galley that turned out first-rate chow for hungry sky sailors and guests aboard. The ship's cook, W.S. Peak, poses with the tools of his trade around the time of the ship's Panama flight (1931).

Pictured in Hangar No. 1 in December 1928, a Vought UO-1 observation plane, fitted with the prototype navy skyhook, is undergoing hangar tests with the new trapeze installed at Frame No. 100 below the keel of the *Los Angeles*. Note the hinging mechanisms and the airship's No. 3 engine car.

A German Prufling glider was carried aloft and sailed down for a demonstration by Lt. T.G.W. "Tex" Settle on Memorial Day 1930 at Naval Air Station, Anacostia. Poor thermal conditions nearly caused Settle to crash in the courtyard of St. Elizabeth's Mental Hospital. Conceived as an idea to land an officer to supervise mooring operations, the glider was soon discarded.

By 1930, "hook-on" flying had become routine. In this view, a Consolidated N2Y-1 fleet trainer makes an approach to the *Los Angeles*. Note the trapeze latticework, the wind-driven fuel transfer pump, and the adjustable fish-mouth radiator opening at the front of the No. 3 engine gondola, with the water-recovery unit just above.

The *Los Angeles* is under overhaul at Lakehurst in the fall of 1929, with sections of cotton outer covering being replaced. Note the overhead tackle suspensions, wooden cradles supporting the gondolas, the front taxiing car under the nose area, and the long fire ladders used by riggers for working on the outside of the giant airship hull. Bosun's chairs, hanging from the hangar rafters, were also used.

The *Los Angeles* is pictured here on the new "iron horse" mobile mast in 1930. Developed and refined by Lt. Comdr. Charles E. Rosendahl in his new capacity of commander of the Rigid Airship Training and Experimental Squadron, the contraption eventually developed into a significant innovation in safety, reliability, and laborsaving.

This is a classic 1929 photograph of a distinguished dirigible skipper. Lt. Comdr. Herbert V. Wiley is at the *Los Angeles* portside control car window, with megaphone in hand. In 1933, as executive officer of the *Akron*, Wiley was one of only three survivors. On February 12, 1935, he was in command of the *Macon* when it was forced to ditch in the Pacific due to structural failure. Wiley spent most of the rest of his career at sea, including command of the battleship *West Virginia* (BB-48) during World War II.

INTERIOR OF HANGAR, NAVAL AIR STATION, LAKEHURST, N.J.

The *Los Angeles*, J-class blimps, and the new ZMC-2 (the "Tin Ship") are at home in Hangar No. 1 in 1930. Naval Air Station, Lakehurst and navy lighter-than-air technology were, by then, moving fully and purposely forward as men were trained and new techniques were developed for the next generation of rigid airships, the ZRS-4/ZRS-5 airplane-carrying naval scout dirigibles.

Station Marines are shown in full dress in the summer of 1931 as the *Los Angeles* makes ready to receive the king and queen of Siam for a short flight. Note the odd experimental water-recovery apparatus above the No. 3 engine gondola.

This photograph was taken on Navy Day, October 27, 1930. The entire lighter-than-air inventory of Naval Air Station, Lakehurst was on parade, as director Frank Capra's camera crews captured footage for the upcoming Jack Holt–Fay Wray film *Dirigible*. Note the kite balloon, free balloons, J-ships, the "Tin Ship," and the star of the show, the *Los Angeles*, on the tripod mast in the background.

Station officers and enlisted men pose with a kite balloon in 1930. Clumsy and obsolete, the hydrogen-inflated contraption remained part of the student lighter-than-air curriculum through the 1930s. Most pilots felt that the free balloons gave much greater feel and experience for fledgling students of lighter-than-air flight techniques.

The Metalclad ZMC-2, built by the Aircraft Development Corporation in Detroit, arrived at Lakehurst in October 1929. Although the design was never repeated, it demonstrated the feasibility and durability a metal-hulled airship concept and survived until 1941. One naval architect even hailed ZMC-2 as "the ultimate airship." Conventional blimps, however, were cheaper and easier to build.

This is the decommissioning crew of the *Los Angeles* on June 30, 1932. Depression-era economy measures forced the ship's retirement after eight years, 331 flights, and 4,500 hours in the air. The navy's most successful rigid airship, the *Los Angeles* trained scores of men for lighter-than-air duty. Although it never flew again, the old Zeppelin survived as a grounded mooring test ship and "hangar queen" until 1939.

With no hope of congressional approval for a modern training blimp, Lakehurst and BuAer personnel created an airship by ordering a "universal" control car for experiments with the J-class ships and a 320,000-cubic-foot envelope with a 51,000-cubic-foot inner cell for "gaseous fuel experiments." K-1 was an odd creation, but it did fly. Shown here as delivered in 1931, K-1 lasted until 1940, when excessive helium leakage and cranky engines caused its retirement.

46

J-3 and J-4 were erected at Lakehurst in 1926–1927. Based on an army blimp design, they were stable, popular training ships. J-3, with its open gondola, was lost in a crash landing at Beach Haven on April 4, 1933, while out searching for the *Akron* survivors. J-4, with its boatlike gondola, flew until 1940. Used for a variety of training and experimental duties, J-4 was especially at home in water landings on Barnegat Bay.

Not even the size of one of the *Los Angeles*'s largest gas cells, the J-ships, with a 196-foot length and a 210,000-cubic-foot helium capacity, could still occasionally give the ground crews a bad time. Here, J-3 is gathered together after being damaged in a landing accident south of Hangar No. 1. Often, the remaining helium could be recovered and reused after the ship was carefully floated back into the hangar.

Here, the J-3 is erected in Hangar No. 1. The big inflation net kept everything in check while helium was slowly admitted through a sleeve in the ship's belly via a gas main leading from the helium plant.

Gregarious Lt. Charlie Bauch is on the hangar deck with his beloved ships in the background. A top-notch airship officer with experience going back to World War I, Bauch was serving as a naval inspector, overseeing construction of the *Akron* when he was tragically killed in an automobile accident in 1931.

Chief Aviation Pilot Frank Masters poses on the Lakehurst field in the garb of a well-dressed lighter-than-air flier. In addition to being occasionally dangerous, duty in naval aviation in those days could be downright cold!

50

Four

THE NAVY SUPER-AIRSHIPS OF THE 1930S

In 1924, some of Germany's best airship designers relocated to Akron, Ohio. By 1929, the Goodyear-Zeppelin Corporation's new "Airdock" was a beehive of activity as Congress authorized $8 million (a significant amount during the Great Depression) for the construction of two giant navy airships, ZRS-4 and ZRS-5. They would emerge in 1931 and 1933 as the *Akron* and the *Macon*.

The *Akron* is moored at Lakehurst on November 2, 1931. The locomotive-hauled railroad mast was capable of hauling the airship in and out of the hangar with half the ground crew previously necessary. The lower fin was made fast to a 133-ton stern beam, and a gas-electric locomotive hauled the stern around against the wind for docking tail-first in Hangar No. 1. Although cumbersome, it was an effective means of handling a 785-foot-long giant.

Adm. William A. Moffett, three-time chief of the U.S. Navy Bureau of Aeronautics, is pictured with sons George Moffett (left) and William Moffett Jr. (right) on a visit to Lakehurst in 1931. A founding father of naval aviation, Admiral Moffett had something of a weakness for publicity. Over the years, many Lakehurst officers quietly grumbled about their beloved admiral forcing them to use their airships for "ballyhoo" and "hand waving" flights instead of serious work with the fleet. A staunch proponent of lighter-than-air technology, Moffett died in his own creation when he was lost with the *Akron* on April 4, 1933.

The *Akron's* lower fin is shown on the stern beam. On February 22, 1932, a freak gust of wind tore the tail loose, right in front of a crowd of congressmen who were coming aboard for a ride as guests of Adm. William A. Moffett. This highly embarrassing mishap put the *Akron* out of service for three months. Dozens of damaged girders had to be replaced, and the lower fin was removed and straightened in draw benches.

The *Akron* flew cross-country in May 1932 to test mooring facilities on the West Coast and operate with the fleet. The trip was a public relations disaster. The *Akron* performed poorly in fleet maneuvers and carried up three men on its handling lines during one landing, two of them falling to their deaths before horrified spectators and newsreel cameramen.

Amidst the 1932 West Coast flight, failures in the water-recovery units caused the sides of the *Akron* to become streaked with black exhaust carbon. The use of makeshift scaffolding, Castile soap, and hand brushes to clean carbon off the ship was an unpopular duty and was considered a form of punishment.

Lt. Comdr. Charles E. Rosendahl (right) hands the *Akron* over to Comdr. Alger H. Dresel on June 22, 1932. Rosendahl saw the big dirigibles as the centerpiece of naval aviation and was never fully comfortable with the "airship-airplane" concept. Dresel, a newcomer to lighter-than-air but an excellent shiphandler, was more open minded to the ideas of the *Akron's* hook-on pilots, allowing them to develop greatly improved tactics. Preferring to do most of his flying Mondays through Thursdays, Dresel was very popular with the crew.

Flying from their tiny hangar in the sky necessitated repeated drills to speed up launching and recovery techniques for the hook-on planes. Squadron aircraft of the *Akron-Macon* heavier-than-air unit consisted of Curtiss F9C-2 Sparrowhawk fighters, conveniently small enough to fit through the T-shaped entrance door in the bottom of the ships' airplane compartment.

U.S.S. AKRON OVER
COUNTY CAUSEWAY
BISCAYNE BAY.

Through the summer of 1932 into early 1933, the *Akron's* crew honed and improved their level of training and efficiency. On January 3, 1933, Comdr. Frank C. McCord took command. It was his ambition to have the *Akron* performing like a true aircraft carrier of the skies.

The *Akron* was intended to embody every important safety feature available, based on eight years of rigid airship flying in the U.S. Navy. With full auxiliary controls located in the giant lower fin, nonflammable helium inflation, internally mounted engines, a heavy "deep ring" main structure, and a host of other improvements, the *Akron* (at least on paper) was the safest, best-engineered dirigible yet built. However, life preservers and rafts were apparently removed from the ship before *Akron's* cross-country West Coast flight and were never put back when it returned to Lakehurst.

Chief Radioman Robert W. Copeland mans his station aboard the *Los Angeles*. The original German-made Telefunken radiotelephone set was augmented by American VHF and RDF equipment. Copeland was lost in the crash of the *Akron* on April 4, 1933.

The loss of the *Akron* at sea was indeed the beginning of the end for the navy's rigid airships. Of the 76 aboard that stormy night, only 3 came back. The next morning, the tragedy was compounded when nonrigid J-3 crashed in the surf at Beach Haven while out looking for survivors, and another 2 men were lost. Recovered *Akron* wreckage was laid out at Lakehurst for inspection by shocked personnel.

The *Macon* is shown under construction at the Goodyear-Zeppelin Airdock in the fall of 1932. One of the fins is shown completed, about to be bolted on. This particular feature of the *Akron-Macon* caused criticism from the Zeppelin Company because German Zeppelins used cruciform girders passing right through the main rings of their airships from one side to the other, and the Germans felt this offered "superior strength." Goodyear's Dr. Arnstein argued that the "deep" structural rings of his *Akron-Macon* design made their tail fins equal to (if not stronger than) the German arrangement. Although the *Akron* probably crashed into the ocean due to a faulty altimeter reading, the *Macon* was definitely lost due to fin failure.

This Martin T4M torpedo plane was stationed at Lakehurst for several years during the 1930s. At one point, station officers wanted to put a skyhook on it for use as an airship refueling tanker, but nothing ever came of it.

Anton Heinen managed to squirrel away some of the lucrative pay from his *Shenandoah* flight instructor days and invested in his own airship product, the Heinen Air Yacht. Shown at Lakehurst's "nonrigid hangar" (later Hangar No. 4), the little hydrogen-inflated ship also used a navy hangar at Cape May in 1933–1934. Heinen's venture and the American Airship Corporation of Lakewood eventually went bust, and the ship was dismantled.

Building No. 120 (left), a modern consolidated services building and barracks, was erected in 1932. Building No. 150 (right) opened in 1942 as headquarters for the chief of the Naval Airship Training and Experimentation Command (CNATE). To the far left, "temporary" wooden buildings housing the Navy Exchange Theater, a bowling alley, and a snack bar remained in use until the 1980s. Note the original one-story administration building in the foreground.

The *Macon* was based at Lakehurst from June through October 1933, but it was felt that if the rigid airship was to prove itself, it would have to spend more time with the fleet. Sent to California and based at Sunnyvale, the *Macon* never returned to Lakehurst again.

The *Akron* was not carrying any planes the night it was lost, and all were transferred to the new *Macon*. F9C-2 No. 9057 proudly displays its ornate striping and "men on the flying trapeze" insignia. (The commanding officer of the Pacific Fleet Air Battle Force was rather miffed by the gaudy paint jobs and slapped the squadron's wrists a bit by ordering them to paint their empennage black as a "color code.")

Good morale, "lifejackets for all," and some of the most experienced men in the navy's lighter-than-air program highlighted the *Macon's* career. In this view, Aviation Chief Machinist's Mate Charles Solar puts in a call to the bridge from one of the ship's port keel phone stations. With lighter-than-air service going back to the "Howden Detachment," the ZR-2, the *Shenandoah*, the *Los Angeles*, and the *Akron*, Solar had more rigid airship flight time than just about anybody else in the navy.

The *Macon* is pictured in October 1933 at Naval Air Station, Sunnyvale (later Moffett Field), in California. Structural weakness in the area of the ship's tail fins was detected in May 1934 Carribean maneuvers, but repair-and-reinforcement work went slowly. The upper fin broke loose and flew apart in a gust of wind off Point Sur, California, on February 12, 1935. Resulting structural damage caused the *Macon* to make a forced landing in the Pacific after losing control and buoyancy. Fortunately, the water was warm, there was plenty of lifesaving gear, and rescue ships were nearby. Of the 83 aboard, only 2 men were lost, but the rigid airship program in the U.S. Navy was finished.

The *Los Angeles* is shown out of commission at Lakehurst in 1935. Note the fabric replacement panels at the bow section, parts of experimental chrome yellow "nose star" still visible, and blimp K-1, which is being towed by a "belly mast."

The former advertising blimp *Defender* was purchased from Goodyear in 1935. Fitted with a new 183,000-cubic-foot envelope and designated G-1, the 192-foot-long ship was very popular, with its handy size, good streamlining, and modern enclosed car. In this view, about 50 men form a ground crew south of Hangar No. 1. Note the airplane hangar to left and the "nonrigid hangar" to the right.

Five

GERMAN ZEPPELINS AT LAKEHURST

With two rigid airships inside, even Hangar No. 1 could get crowded. In this October 1928 photograph, Dr. Hugo Eckener has arrived from Germany with LZ-127 *Graf Zeppelin* on its first transatlantic voyage. Seen from the east end, the *Graf Zeppelin* is having damaged fabric on its port stabilizer replaced (foreground). The *Los Angeles*—with blimps J-3 and J-4 tucked under its tail and a new mooring mast being assembled under its nose—is virtually immobilized.

The *Graf Zeppelin* is pictured during one of its four landings at Lakehurst. Lakehurst was then appearing in headlines around the world as the departure and destination points for the *Graf Zeppelin*'s 1928 transatlantic flight and 1929 round-the-world voyage as Dr. Hugo Eckener and his seasoned crew amply demonstrated the superiority of the big dirigible for long-distance travel.

At Mines Field in Los Angeles (later the site of LAX), U.S. Navy airship officer Lt. T.G.W. "Tex" Settle moves a little too quickly for the camera as he assists in fueling, gassing, and loading the *Graf Zeppelin* for its 1929 round-the-world voyage. The long gondola featured a "bridge" area forward, passenger accommodation at rear. Note the retractable wind-driven generator beneath chart room window and bulky rattan "air bumper" for cushion in rough landings.

Intrepid journalist Lady Grace Drummond-Hay caught the attention of the world with her brilliant reporting from the sky, sent to the Hearst Press via the *Graf Zeppelin's* powerful radio transmitters. Aboard for many close calls on the early flights of the ship, her steady nerves and disposition earned her the respect of seasoned crewmen and fellow male travelers alike. (She perished shortly after World War II, following imprisonment under Japanese occupation in the Far East.)

In addition to several tons of mail and freight, the *Graf Zeppelin*, flying along at 70 miles per hour, had berths for up to 20 passengers, gourmet meals freshly prepared in an all-electric galley, and an amply furnished dining-lounge area that offered far greater creature comforts than any contemporary airplane cabin. Although smoking was not permitted due to the use of hydrogen gas for lift, there was, at least, the finest wine list ever to sail the skies.

The *Graf Zeppelin* poses for the newsreel cameras through the west doors of Hangar No. 1 in 1929. Until grounded in June 1937 (after the *Hindenburg* disaster), the *Graf Zeppelin* carried more than 13,000 passengers and flew more than one million miles in 590 flights.

By 1933, Hugo Eckener and the *Graf Zeppelin* had established regular passenger, mail, and freight business between Germany and South America. In a big new building hangar at the Zeppelin works in Friedrichshafen, the Zeppelin Company was ready to build its "ideal" commercial airship.

With a large infusion of cash from the new Nazi government in Berlin, construction of the LZ-129 proceeded at Friedrichshafen in 1934–1935. Note the integral cruciform tail structure passing through the ship from one side to the other. From this point forward, Berlin had quite a bit of say in operation of the airships. By late 1933, the *Graf Zeppelin* had swastikas painted on its tail fins, as did the new LZ-129 when it emerged in 1936 as the *Hindenburg*. At this time, the government-sponsored German Zeppelin Transport Company would handle all Zeppelin flight operations.

Friedrichshafen a. B. mit LZ 129 „Hindenburg" vom Flugzeug aus

This 1936 photograph shows the Zeppelin Company building works at Friedrichshafen. With both the *Hindenburg* and the *Graf Zeppelin* shuttling between North and South America and the new LZ-130 under construction (inside the building hangar at the right), German Zeppelin Transport Company personnel looked ahead to a truly global service, with 40 or 50 airships linking cities of the globe by 1945. With the growing confidence (and chauvinism), they proposed a joint German-American airship line, offering to supply their superior designs and operational know-how in exchange for nonflammable American helium.

Despite anti-Nazi sentiment growing within the United States (and his own feuding with officials in Berlin), Dr. Hugo Eckener was able to secure a permit through Pres. Franklin Roosevelt for use of Lakehurst in 10 round-trip demonstration flights across the North Atlantic for the 1936 season. On the first arrival (May 9, 1936), the old, decommissioned *Los Angeles* was on the field to greet the new German arrival.

The *Hindenburg* is shown moored to the Wellman railroad mooring mast at Lakehurst in 1936. Note the windows of the passenger quarters along the sides of the ship and the Olympic rings commemorating the games held in Berlin that summer. Crossings became commonplace, with the Zeppelin being landed, serviced, and turned around for Germany in six to seven hours. The Germans took great pride in prompt arrivals and departures, making much ado of the times when American Airlines's DC-3 connecting flights from Chicago and Newark arrived late, delaying the *Hindenburg*'s takeoff.

The *Hindenburg's* lower fin is secured to a riding-out car on the Lakehurst mooring circle in 1936. While the base commanding officer, Comdr. Charles E. Rosendahl, was able to provide navy bluejackets as part of their lighter-than-air training, the German Zeppelin Transport Company had to supply the balance of the landing crew by hiring men and boys from town at a rate of $1 per hour.

With cabin space for 50 passengers (increased to 72 for the 1937 season), the *Hindenburg* was accepted as safe, practical transportation, despite the use of hydrogen gas. Transatlantic fare was $400 for a one-way ticket and $720 for a round-trip ticket. Crossings averaged 61 hours westbound and 52 hours eastbound. Passenger cabins were a bit cramped, but the lounge, dining room, writing room, fireproof smoking room, and nearly round-the-clock bar service offered amenities similar to those on an ocean liner.

The *Hindenburg* is shown over lower Manhattan in 1936. Although the giant Zeppelin later became a long-recognized symbol of tragedy, events 65 years later at this location redefined the term.

The *Hindenburg* is moored to the Lakehurst stern beam. Note the four-bladed propellers and the massive fins. Soft sand ultimately cushioned the fall of many survivors who were to jump from the burning airship a year later.

To the envy and chagrin of many U.S. Navy lighter-than-air personnel, the *Hindenburg* established regular, reliable service across the Atlantic in 1936, carrying more than 1,000 passengers, mail, and freight items ranging from tiny packages to a disassembled biplane and this Opel automobile.

Capt. Ernst Lehmann (fourth from left) poses with a group of happy *Hindenburg* passengers in 1936. With Dr. Hugo Eckener now regularly feuding with Nazi officials in Berlin, Lehmann enjoyed rubbing elbows with the New Order and soon became director of flight operations. The charming, dapper "little captain" died of burns in the *Hindenburg* disaster the following year.

The *Hindenburg* is aloft in March 1937. Olympic ring markings have been painted over, more passenger cabins added, and the Germans are even experimenting with hook-on airplane to facilitate exchange of passengers and mail from the ground. For the 1937 season, 17 round trips were planned between Frankfurt and Lakehurst, with a few South American trips sandwiched in between. The *Graf Zeppelin* continued to fly the bulk of the South American flights that year, with the new LZ-130 scheduled to enter service on October 27.

All hopes and dreams for the future were abruptly shattered on May 6, 1937, when the *Hindenburg* caught fire and burned while mooring at Lakehurst on arrival from its first North Atlantic crossing of the season. In 34 seconds, the ship crashed into a burning, smoking heap on the field west of Hangar No. 1. There were 97 people aboard the ship, and 13 passengers, 22 crewmen, and 1 civilian member of the ground handling party were killed.

The burned-out skeleton of the *Hindenburg*'s nose section is pictured here. Dismantling took place over the summer months, and many souvenirs were had by the locals. The cause of the disaster is still debated today. Theories range from static electricity of a passing thunderstorm to a time bomb to the Germans' unwitting use of inflammable "dope" on the outer fabric. Some even entertain stories of stray model rockets, concealed snipers, and the reckless use of gasoline to prime the ship's diesel engines.

Silverware, cups, and a serving tray from the *Hindenburg* wreckage are part of the extensive collection of the Navy Lakehurst Historical Society. Dr. Hugo Eckener hoped to persuade American authorities to relax their export ban on helium so he could resume passenger service with LZ-130 by 1938, but German aggression in Europe made such prospects hopeless.

The brief, exciting chapter of the commercial Zeppelins ended with the *Hindenburg*. Within a few years, Lakehurst's role would transform from welcoming German Zeppelins to sending out patrol blimps in search of German submarines. War was again on the horizon.

Six

NAVY LAKEHURST IN WORLD WAR II

Comdr. Charles E. Rosendahl (center window) accompanies the crew of J-4 for a short hop. Sometimes, these were for training or experimental flights. Sometimes, the ship flew just to make sure that those aboard were getting their required four flying hours per month to qualify them for the additional 50 percent flight pay.

Sometimes, it just does not pay to get up in the morning! In this undated photograph, G-1 is "ripped" after its tail has caught the doors of Hangar No. 1. Such accidents, while embarrassing, were seldom serious, and the ship could be repaired and reinflated in a week.

Training blimp L-1, identical to the standard six-passenger advertising blimps operated by Goodyear, is pictured outside the east doors of Hangar No. 1 in the summer of 1939. The *Los Angeles* is inside, shortly before it disappeared forever. As the curtain came down once and for all on the rigid airships, the L-ships' careers were just beginning. Over the next several years, the navy operated 22 of them for training and limited patrol work.

G-1 undocks through the east doors of Hangar No. 1 in 1939. Note the tractor-drawn mobile mooring mast and wooden loft of the Parachute Rigger School over the shops along the interior south wall. Lakehurst's first bad casualty of World War II, G-1 collided with an L-2 off the Manasquan Inlet during secret nighttime tests of a high-intensity underwater flare. Both blimps ripped open on impact and crashed into the ocean with 12 men killed on June 8, 1942.

Big patrol blimps TC-13 and TC-14 transferred from the U.S. Army Air Corps after the army terminated its blimp operations in 1937. As good as anything the navy had, these relatively modern ships formed the nucleus of the first patrol squadrons developing antisubmarine warfare tactics at Lakehurst by 1940. Shipped to the West Coast in 1942, the war-weary TC-14 eventually ended up back at Lakehurst, tied to a stick mast to field-test water-emulsified Johnson wax for use as a blimp envelope snow repellent.

K-2, a $170,000 prototype of a new generation of U.S. Navy airships, arrived at Lakehurst on December 16, 1938. With a little modification, the K-type ships—fast, long-ranged, and capable of staying aloft 40 hours on patrol—became the standard navy patrol blimps of World War II. Over the next six years, 133 similar sisters were delivered. Despite later complaints from certain circles about monopoly and "price gouging," Goodyear remained the navy's sole supplier of airships until the end of the program.

This is the graduating class of the Lakehurst lighter-than-air school in May 1942. Note how the building windows are taped in case of air raids. The chief petty officer (later a lieutenant commander) standing front and center is L.E. "Gene" Schellberg, a veteran from the days of the *Shenandoah*. By 1943, another veteran lighter-than-air man, Rear Adm. Charles E. Rosendahl, would hoist his flag as chief of the Naval Airship Training and Experimentation Command, while Capt. Raymond F. Tyler would become commander of Fleet Airship Wing 1.

An early K-ship (probably K-4) passes over a torpedoed tanker off the New Jersey coast in June 1942. The blimp's prewar striping and markings have been painted over for "security reasons." The few K-ships available at this stage of the war were stretched to the limits of their range of endurance. Within a year, more blimps over shipping lanes and better-organized convoy systems drastically cut shipping losses to the prowling German U-Boats.

This K-ship is tied down in Hangar No. 1. The air duct passing through the car door enables the crew on the "pressure watch" to maintain ballonet pressure with portable blowers. Note the 425-horsepower Pratt & Whitney engine, sandbags on the handrail for ballast, and the swiveling landing wheel with which the blimps could take off "heavy," like an airplane.

WAVES assumed the duties of "pigeonaires" at wartime Lakehurst. Each blimp on patrol carried up to six homing birds to assist in sending messages while maintaining radio silence.

The Brazilian air force delegation is shown with K-ship officers in Brazil in 1944. Blimp Squadron ZP-41 used the former German Zeppelin hangar at Rio in support of lighter-than-air operations over the South Atlantic. Brazil was interested in a lighter-than-air program of its own, but the training program set into motion was cancelled by Brazilian authorities right after V-E Day.

K-ships abound on the field west of Hangar No. 1 in 1943. As headquarters of navy lighter-than-air activity in the Eastern Sea Frontier, Lakehurst grew beyond the wildest expectations of prewar planning. With the rigid airships gone, blimps were proving the worth of lighter-than-air vehicles in antisubmarine warfare, convoy escort, and air-sea rescue missions.

In further attempts to "live with the fleet," blimps were landing on carrier decks to refuel and rearm by 1945. Here, off California, K-29 prepares to touch down on the deck of the *Altamaha*. Such procedures became routine and greatly enhanced the range and utility of the modern patrol blimps.

Steel Hangars No. 2 and No. 3 were under construction by 1941. Hangar No. 2 initially housed Lakehurst's assembly-and-repair department. Hangar No. 3, slightly shorter in length, housed the smaller training blimps. Today, these hangars house Manufacturing and Prototyping, which provides service to the fleet.

As a result of wartime steel shortages, Hangars No. 5 and No. 6 were among the largest wooden structures in the world. Seventeen wooden-arch hangars were erected at naval air stations on the Atlantic, Pacific, and Gulf coasts. Completed in 1943–1944 as wartime expedients, seven of these 1,175-foot sheds (including the two at Lakehurst) are still standing today.

With the establishment of the Parachute Rigger School in September 1924, Lakehurst's role expanded beyond airships and into other areas of the naval establishment. With the base's airship activities drastically cut back from 1934 to 1939, the Parachute Rigger and Aerographer's Mate (weather) Schools helped keep Lakehurst's hand in the mainstream of naval aviation. "Temporary" wooden Building No. 190 was added during huge World War II expansion in 1942 to house the Parachute Rigger School.

This K-ship, with elevators up, is pictured over West Field while another stands off to the north. With as many as 27 blimps operating at any one time, Lakehurst relied on a control tower and regular air-traffic control procedures for flight operations.

M-ship XM-1 is carrying Piper Cub *Glimpy* for drop tests in 1945. Although not a hook-on plane in the sense of the *Akron-Macon* era, the *Glimpy* was designed to ferry observers, classified films, documents, or rescue-related personnel and material quickly ashore. This M-ship eventually went on to perform a record 170-hour endurance flight in the summer of 1946, the first of many postwar records set by navy blimps.

Helium-inflated training balloons pose adjacent to Mat No. 1 in 1944 for an MGM crew filming scenes for *This Man's Navy*, Hollywood's salute to the navy blimps of World War II. The film starred Wallace Beery, Tom Drake, James Gleason, and Jan Clayton.

L-19, one of the L-type training blimps developed by Goodyear, was based on the company's late-1930s advertising blimps. Small, handy, under 150 feet in length, and about a quarter of the cubic volume of the standard K-type patrol blimps, the L-ships were usually assigned to training and utility work during the war, mostly on the West Coast. After serving most of its navy career at Lakehurst, L-19 was purchased as war surplus for $5,000 and enjoyed a long career as a commercial advertising airship in Europe and Japan until wrecked by a typhoon in 1969.

"They were dependable" is how Adm. Charles E. Rosendahl best described the performance of his beloved airships in World War II. The claim certainly holds some merit; during the war, blimps safely escorted 89,000 surface ships in all types of weather, around the clock, often when airplanes were grounded. Only one blimp is confirmed lost to enemy action, but one-fifth of the navy's blimps were damaged, ripped, or wrecked beyond repair in operational mishaps. Of equal frustration was the fact that they never got credit for sinking a submarine.

Seven

THE COLD WAR

Lakehurst, New Jersey

In this view, a K-ship from the Naval Air Reserve Training Unit (NARTU) is readied for flight on the mat to the east of wooden-arch Hangar No. 5.

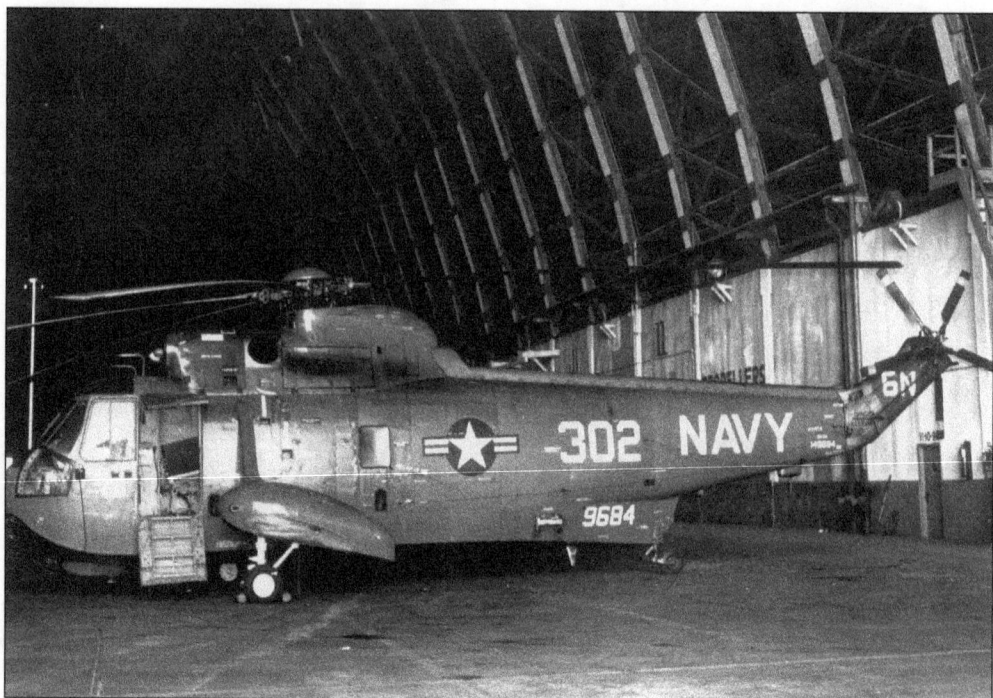

The SH-3A helicopter assigned to Reserve Squadron HS-751 takes the place of blimps as it occupies floor space in Hangar No. 6.

This aerial view of the Naval Air Test Facility, Ship Installations (NATF-SI) shows the steam catapults, the two-and-a-half-mile runway, the powerhouse, and the associated support buildings.

The graduating class of the Parachute Rigger School is assembled on the steps of Building No. 190 in January 1952. Note the soon-to-be-discontinued "Donald Duck" hats.

As part of his final qualification as a U.S. Navy and Marine parachute rigger, the student was required to make a high-altitude jump with a self-packed parachute. This instilled the responsibility of their task like nothing else; most trainees would make up to five jumps before graduation. Here, a group goes aboard one of the Station R4D utility aircraft for a practice jump flight.

The first fleet helicopter squadrons were formed and developed at Naval Air Station, Lakehurst in 1947. These pioneering craft of HX/VX-3 developed techniques for eventual fleet deployment, and their subsequent success in search-rescue and antisubmarine warfare duties contributed to the demise of lighter-than-air operations within 15 years. This is an early Bell HTE-1 training/light observation helicopter.

With the onset of World War II, the base perimeter was relocated to the east boundaries of the field. A "temporary" wooden security building survived for decades, to be replaced in the late 1980s by a modern pass-identification structure.

By the early 1950s, the U.S. Navy deployed the Piasecki HUP Retriever (above) for air-sea rescue and fleet support. On Lakehurst Mat No. 1 (below), a Sikorski HO4-S utility helicopter of Squadron HC-2 stands ready as an early Bell HTL lifts off on a training hop.

Lakehurst featured a wide variety of "hack" aircraft over the decades. The Beechcraft C-45 Bugsmasher was assigned to the base in the 1960s and 1970s for utility and training purposes.

Navy R4D, the military version of the famous Douglas DC-3, was assigned to Lakehurst for transport and parachute training.

As a recruiting incentive, young prospective U.S. Naval Reserve aviators were given the opportunity to experience the thrill of flight firsthand with a ride in Naval Air Reserve Training Unit's T-34 Mentor. These venerable "taxi" planes operated at Naval Air Station, Lakehurst through the 1970s.

Pictured c. 1972 is an SH-2 Sea Sprite utility helicopter from Squadron HC-2 above the mat just east of wooden-arch Hangar No. 5 (right).

The year 1970 saw the formation of Carrier Air Reserve Wing 70, and Naval Air Station, Lakehurst hosted the majority of the air wing's squadrons, including VS-71, VS-73, and HS-75. Such aircraft as this Grumman S2-E Tracker were a common sight as they flew their tireless antisubmarine warfare patrols over the Jersey Shore.

Grumman S-2E of VS-73, "the Fighting Blue Bandits," awaits a midafternoon training flight. While on a 1973 two-week active duty deployment to Naval Air Station, Quonset Point, all VS-73 aircraft and assets were transferred to Naval Air Station, Detroit, making it one of the "lost squadrons" of Lakehurst.

Base Marines are shown at "morning colors" outside Building No. 120 in 1969. The flagpole area still features the ship's bell, and the Browning .50-caliber guns came from old battleship *Maine* (BB-10). The guns were an early gift to the base from Adm. Robley C. Evans, whose son Capt. Frank T. Evans was the first Lakehurst commanding officer. Marines remained stationed at Lakehurst through 1985.

Built in 1932 with donations from the American Legion and other civic organizations, the Cathedral of the Air is located outside the area of the main gate, near the town of Lakehurst. The multidenominational chapel continues its service to navy and civilian worshipers from nearby areas.

This photograph was taken during an open house in 1956. One of the new ZS2G-1 antisubmarine warfare patrol blimps is hovering low. To the right is a Tilley aircraft crane, with the floating-roof helium-storage gasometer for the Lakehurst helium plant in the background.

A ZS2G-1 airship, its propeller blades visibly damaged, approaches a crash crew ready for recovery after the ship brushed trees in a botched approach. While a blimp had an advantage in that it could float, the crew was probably pretty well shaken up anyway.

ZPG-2W, an all weather airborne warning (AEW) variant of the "Nan" ship, cruises east of the base near Pine Lake Park c. 1957. With a height-finder radome and a huge antenna revolving around inside the helium bag, these and the larger ZPG-3Ws were conceived as seaward extensions of the nation's distant early warning (DEW) line but were shunted aside before full deployment.

N-type, or "Nan," ships are pictured on the mat at West Field on a bitter, crisp morning. Although airship mooring had become comparatively refined by the 1950s, the weight of wet snow was always a problem. Mast watch under such conditions was cold and arduous. Note the helium truck near the mast.

By 1961, when termination of lighter-than-air was imminent, Naval Air Station, Lakehurst was the only base in the fleet offering overhaul-and-repair service to heavier-than-air, rotary-winged, and lighter-than-air aircraft. Despite this versatility, Lakehurst's overhaul-and-repair department was eventually phased out along with the lighter-than-air program. WV-3 Warningstars await reworking outside overhaul-and-repair hangars (above) while an early Kaman HOK-1 helo undergoes overhaul inside.

The navy's last four blimps were the giant ZPG-3W type of 1.5 million cubic feet. Delivered between 1958 and 1960, they remain the largest nonrigid airships built to date. On July 6, 1960, one of the big "3Ws" apparently suffered a catastrophic failure, collapsing in flight and falling into the sea. The blimp quickly sank, and only 3 of the 21 men in the crew got out alive. While Goodyear argued that the ship had been improperly flown and maintained, naval personnel contended that Goodyear had sold the navy a defective airship. This became a moot point, as the navy had only 10 airships in service by January 1961.

Although squadron activity terminated in 1961, two ZPG-2 blimps remained in research-and-development service through August 1962. One assisted in Operation Clinker (a classified antisubmarine warfare infrared-detection experiment), while the other (foreground) engaged in "flying wind tunnel" activities in conjunction with Princeton University. The end is near as the last two navy blimps repose in Hangar No. 1 in the summer of 1962.

Air station officers are pictured in 1962. With two helicopter utility squadrons, antisubmarine warfare patrol squadron VS-751, and the navy's only aerology and parachute rigger schools, the future of the station seemed reasonably secure as the last blimps were deflated in the fall of 1962. With navy lighter-than-air now officially dead, certain elements of the naval aviation hierarchy took great delight in completely discarding most operational and historical reminders from the lighter-than-air period.

Capt. M. Henry Eppes was the highest-ranking airship officer in June 1961, when it was announced that navy lighter-than-air would be phased out. Retiring from active duty the same day that the last squadrons stood down, Eppes pursued an academic career and, in 1986, spearheaded the founding of the Naval Airship Association. In June 2001, he flew from California to help commemorate Navy Lakehurst's 80th anniversary and cut a special cake honoring his own 89th birthday, which sadly turned out to be his last.

The vast interior of the base fabric shops (Building No. 123) allowed space for the packing and inspection of parachutes, as well as the repair and inflation of blimp envelopes. The building, with its low-abrasion polished maple floor, now serves as a fitness and recreation center.

North American FJ-4B Fury NATF No. 5 is launched from the steam catapult with Lt. Comdr. Bob Ellis at the controls in October 1960. Although an aircraft carrier launched planes while steaming 35 knots into the wind, the aircraft at the Naval Air Test Facility (NATF) were hurtled into the air from a standstill with zero forward speed.

Standing guard in front of the test facility, this 1950s F-8A was restored by members of the Navy Lakehurst Historical Society. The society designed the NAEC logo and performed full cosmetic restoration.

This dramatic view shows a fully loaded F-8E Crusader launched from the Naval Air Test Facility steam catapult. Fitted with a variable-pitch or "tilting" wing, the F-8 was able to attain increased lift and landing performance.

The F3-D Screamin' Demon catches the wire at the Naval Air Test Facility in 1961. These big, all-weather, single-engined interceptors were eventually replaced by the McDonnell-Douglas F-4 Phantom by 1964.

At Defense Secretary Robert McNamara's insistence for a "standard" interceptor aircraft for use by both the navy and air force, the General Dynamics F-111B came to Lakehurst for carrier suitability tests, but the navy resisted and eventually adopted the F-14 Tomcat. The aircraft shown here was severely damaged during barrier crash tests and languished at Lakehurst until the mid-1970s, when it was finally scavenged for parts by the air force.

The Naval Air Test Facility was assigned one of every type of fleet fighter and attack aircraft for test and evaluation, and the naval aviators assigned there got stick time on an impressive variety of models.

A Douglas A4-D Skyraider (also known as a Ford) awaits test launch at the Naval Air Test Facility in 1959. A superb interceptor aircraft, this is one of the only type of naval aircraft to serve in conjunction with NORAD.

This North American RA-5C Vigilante survived many combat missions over North Vietnam and was damaged in tests at the Naval Air Test Facility years later. Cosmetically restored by Naval Air Technical Training Center personnel, this aircraft was destroyed when accidentally dropped over Burlington County by a Delaware National Guard helicopter engaged in transporting it to a new home at the Air Victory Museum. It was a sad end to a proud warrior.

A Vigilante undergoes initial flight tests at Lakehurst in October 1960.

The jet sled tracks have seen a myriad of interesting aviation experiments. In this late-1960s photograph, an ejection test is being performed, closely duplicating an actual combat ejection under controlled conditions. The A-4 cockpit used in the tests is now part of the collection of the Navy Lakehurst Historical Society.

This mid-1970s aerial view of Navy Lakehurst shows the test facility runways and test tracks (upper left), the parachute school drop circle (upper center), the crosswise runways of West Field, and wooden Hangars No. 5 and No. 6 (center). To the lower right are Hangars No. 1, No. 2, and No. 3, with Mat No. 1 (originally Maxfield Field) to the west. Hangar No. 4 is to the extreme lower right.

With a lack of civilian encroachment, Lakehurst, covering 11.5 square miles, was able to host many different activities, as is evident in this photograph of the jet-sled test tracks reaching west toward the boundary with Fort Dix.

In this photograph, the test aircraft has been propelled to crash speed by the jet-sled test tracks, and the dramatic failure and catastrophic ignition of fuel is evident. Note the flying debris as the aircraft hits the test barriers, rupturing the fuel cells and causing the disintegration of the airframe.

An obsolescent Lockheed P-2 Neptune patrol plane is sacrificed for a fuel gel crash test at the Naval Air Test Facility in the early 1970s. The facility would remain a major tenant command through the 1970s, along with the Naval Air Technical Training Center (NATTC) and the Naval Air Engineering Center (NAEC), but Lakehurst's actual days as a commissioned naval air station (the nation's third oldest after Pensacola and North Island) were numbered. On March 13, 1977, Naval Air Station, Lakehurst ceased to exist and was renamed Naval Air Engineering Center, Lakehurst. Today, the base is known as the Naval Air Engineering Station.

Eight

LAKEHURST ODDS AND ENDS

Rising to chief of the Naval Airship Training and Experimentation Command during World War II, Vice Adm. Charles E. Rosendahl retired to nearby Toms River in 1946. An outspoken proponent of the airship cause, he remained a regular presence at Naval Air Station, Lakehurst until his death in May 1977, at age 85. Bitterly disappointed with the navy's decision to remove Lakehurst's designation as an active naval air station, he abandoned plans for establishing a museum at his beloved base, and his accumulated archives instead went to the University of Texas, in his former home state.

As a result of "creative squirreling away," enough components to put a ZPG-2 and a ZPG-3W airship back together actually existed in Lakehurst storage through 1976, when the navy's last two blimp bags were laid out in Hangar No. 5 and air-inflated for evaluation. The big cotton 3W bag, badly rotted, was cut up. The Dacron ZPG-2 bag, in perfect shape, went on to be sacrificed in the Piasecki *Heli-Stat* disaster of 1986, the same year these last two blimp-control cars were photographed next to Hangar No. 6.

In the late 1970s, the Piasecki Aircraft Corporation attracted government interest in its *Heli-Stat,* heavy-lift lighter-than-air, concept. Using the Dacron bag, fins, and components from Lakehurst's last stored ZPG-2 blimp, *Heli-Stat,* years behind schedule and over budget, was the recipient of Sen. William Proxmire's Golden Fleece Award. It shook itself apart and collapsed in a blaze of burning fuel during taxiing tests on July 2, 1986. One crewman was killed.

CALASSES (Carrier Aircraft Launch and Support Systems Equipment Simulator) was erected in the east end of Hangar No. 1 in 1968. A working one-third scale model of a fleet carrier flight deck, it enables hands-on training for aviation bosun's mate ratings, including catapult, arresting gear, and firefighting trainees. For 30 years, every carrier deckhand in the U.S. Navy trained at Lakehurst.

Over the years, various commercial enterprises have used the sprawling Lakehurst hangars for the rigging, inflation, and testing of blimps and aerostats. Here, a tethered aerostat is ground-checked in Hangar No. 5.

The Naval Air Engineering Center, Lakehurst is shown in the mid-1980s. The old East Field in the foreground is now heavily planted in pine trees, with a baseball diamond in front of the long-inactivated east doors of Hangar No. 1. Next to Hangar No. 1, steel Hangars No. 2 and No. 3 now house the precision machining and manufacturing facilities of the Naval Air Engineering Station. Across the field, wooden-arch Hangars No. 5 and No. 6 are used for a variety of aircraft purposes, and beyond them are the parachute jump area and the sprawling runway installations of the Naval Air Test Facility.

While some visitors to Lakehurst jostled to view exhibits or climb aboard display aircraft, others were content to find a good spot to view the aerial demonstrations. This grandmotherly woman has obviously found her own version of the best seat in the house.

Lakehurst's leadership and expertise in aircraft launch and recovery operations resulted in visits by aircraft from programs around the world. Most NATO countries have tested their aircraft at the Naval Air Test Facility, such as this Royal Navy Fleet Air Arm F-4K being readied for launch. Note the British practice of raising the nose for increased lift, necessary because of the Royal Navy's shorter flight decks.

Inside Hangar No. 2 (once the home of Lakehurst's overhaul-and-repair department), Manufacturing and Prototyping serves the needs of carrier aviation in the 21st century. The facility includes computer-assisted design, fabrication, welding, and manufacture of all launch-recovery hardware used in naval aviation today.

One of Lakehurst's many roles today is the design, development, and construction of all the "yellow gear" (aircraft-support equipment), as shown in this impressive array gathered around an F-14 Tomcat.

Over eight decades, Lakehurst has hosted many air demonstrations, highlighting everything from rigid airships and balloons to the Blue Angels. In this photograph, the public views models, pictures, and historical items in a massive 40-table display provided by the Navy Lakehurst Historical Society for the 1984 air show.

The U.S. Navy's newest fighter, the F/A-18F Superhornet makes a landing approach to the Naval Air Engineering Station, Lakehurst, undergoing carrier suitability tests. With more than 80 years of service to the fleet, Lakehurst remains at the cutting edge of naval aviation technology.

One of the test facility's unique features is its fixed shipboard landing platform. In this view, a modern SH-60 Seahawk simulates landing at sea.

As the navy's sole supplier of crash barriers and other shipboard safety equipment, Lakehurst often gets to test its handiwork right in its own backyard, as illustrated by this F-14 Tomcat coming to a stop with the aid of a nylon barrier at the Naval Air Test Facility.

A French navy Raphael fighter from the air group of the carrier *Charles DeGaulle* undergoes shipboard suitability tests at Lakehurst in the spring of 2000.

With the end of lighter-than-air operations, there was serious debate on whether the big hangars should remain, but in their new roles of manufacturing, training, testing, and deployment of modern aircraft, structures dating back more than 60 years continue to serve the needs of modern naval aviation. Modern commercial blimps are still occasionally housed at the former home of the navy's lighter-than-air program.

Harking back over the years, there are still a (very) few people who recall such events as Charles Lindbergh's 1929 inspection visit and the 1940 Navy Ball at the Sea Girt Inn, Sea Girt, New Jersey, with Eddie Cantor as the master of ceremonies.

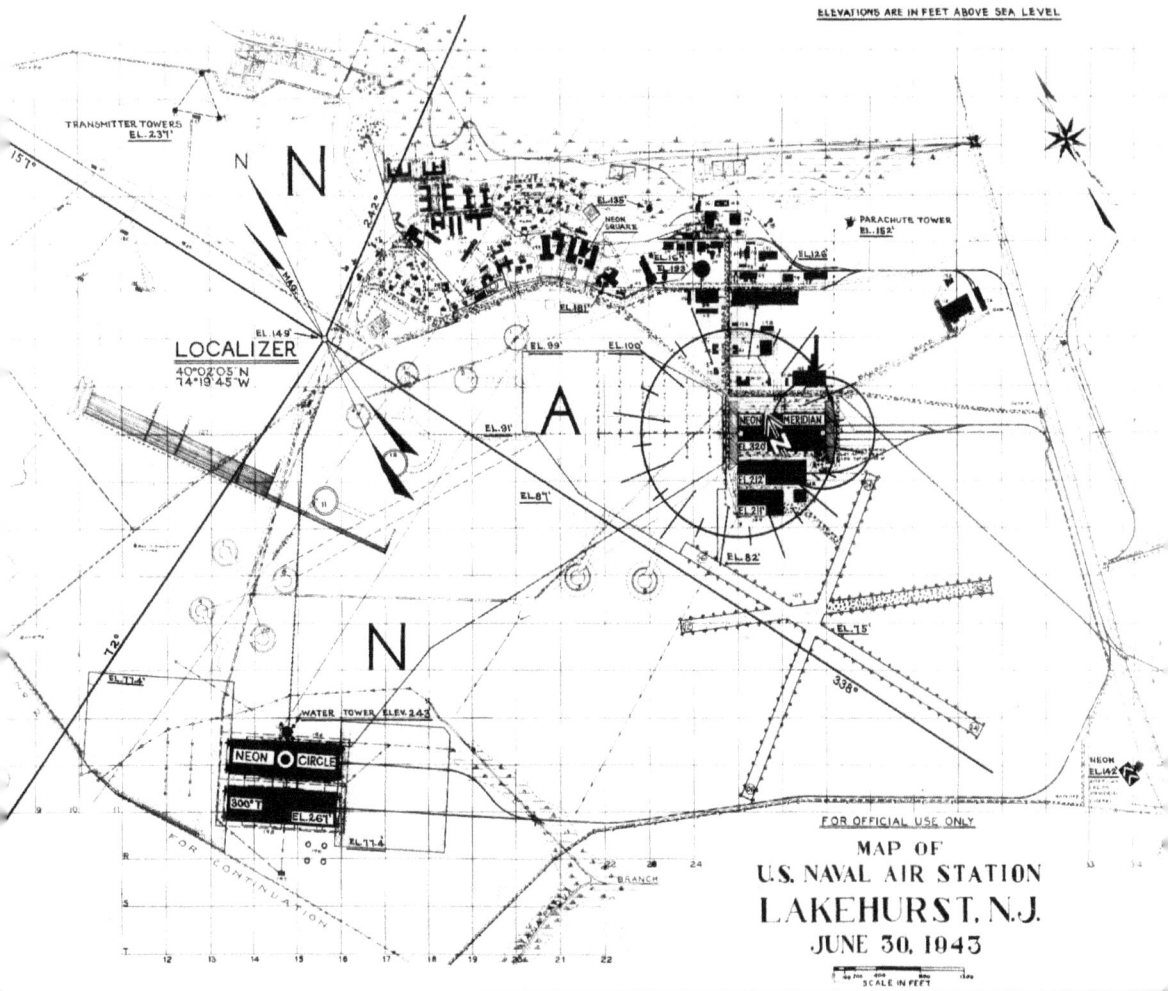

Navy Lakehurst's wartime expansion and layout shows in this 1943 map.

These patches commemorate the different aspects and contributions of Navy Lakehurst over the years.

www.ingramcontent.com/pod-product-compliance
Lightning Source LLC
Chambersburg PA
CBHW050645110426

42813CB00007B/1923